YOUR KNOWLEDGE HAS VALUE

- We will publish your bachelor's and master's thesis, essays and papers

- Your own eBook and book - sold worldwide in all relevant shops

- Earn money with each sale

Upload your text at www.GRIN.com and publish for free

Bibliographic information published by the German National Library:

The German National Library lists this publication in the National Bibliography; detailed bibliographic data are available on the Internet at http://dnb.dnb.de .

This book is copyright material and must not be copied, reproduced, transferred, distributed, leased, licensed or publicly performed or used in any way except as specifically permitted in writing by the publishers, as allowed under the terms and conditions under which it was purchased or as strictly permitted by applicable copyright law. Any unauthorized distribution or use of this text may be a direct infringement of the author s and publisher s rights and those responsible may be liable in law accordingly.

Imprint:

Copyright © 2012 GRIN Verlag, Open Publishing GmbH
Print and binding: Books on Demand GmbH, Norderstedt Germany
ISBN: 9783668315426

This book at GRIN:

http://www.grin.com/en/e-book/341630/power-to-gas-technology-evaluation-of-the-technology-s-potential-in-the

Julien Gianoncelli

Power-to-Gas Technology. Evaluation of the technology's potential in the German energy market

GRIN Publishing

GRIN - Your knowledge has value

Since its foundation in 1998, GRIN has specialized in publishing academic texts by students, college teachers and other academics as e-book and printed book. The website www.grin.com is an ideal platform for presenting term papers, final papers, scientific essays, dissertations and specialist books.

Visit us on the internet:

http://www.grin.com/

http://www.facebook.com/grincom

http://www.twitter.com/grin_com

Seminar Paper

Products and Operations

"Power-to-Gas Technology"
"Evaluation of the technology's potential in the German energy market"

carried out at
Karlshochschule International University Karlsruhe
in the degree course "International Energy Management"

Name: Julien Gianoncelli

Location, date: Karlsruhe, the 30.12.2012

Table of contents

Table of figures

Prologue

The energy efficiency targets of the *"Federal Government"* are important parameters deciding upon the amount of energy we will need in the future and hence about the storage capacities we will require. The reduction of the primary energy consumption by 20% until 2020 and 50% until 2050 towards 2008's figures, the reduction in electricity consumption by 10% until 2020 and 25% until 2050, the doubling of the rate of refurbishment from 1% to 2% per year and the reduction of the end energy consumption in the mobility sector by 10% until 2020 and 40% by 2050 towards 2005's consumption[1] will change the energy market substantially. For instance, houses will need less electricity for their devices and less gas for heating, load profiles will smoothen and passenger and freight traffic will switch from petrol engines to electric and gas engines, fuel cells or will switch to rail.

In Germany, the market structure has been designed to enhance and facilitate the development of renewable energies through the *"Renewable Energies Act"* (EEG) and the *"Energy Economy Law"* (EnWG). And actually, in 2011 the share of renewables in the electricity production totalled to over 20%[2] and is expected to rise over 25% this year, whereby especially the photovoltaic is booming with growth rates of nearly 50%.[3] The evolutions are certainly pleasant, still, the very quick development of these, mostly decentralised and location dependent energy sources, is making the market framework conditions come to its limits. The transmission grid capacities are not large enough to transfer the huge amounts of wind energy from the north to the south and the distribution grid is often too weak to feed in the large amounts of photovoltaic energy in the midday. These restrictions have caused 407 GWh of wind energy to get lost in 2011 (150 GWh in 2010)[4], a small but exploding figure, and a valid argument for storage capacities.

This shows that the whole of the energy market has to be evaluated and diverse scenarios for the future have to be created, to forecast the development into one specific area. Additionally, because of the dynamic of this market, these analyses and forecasts have to be updated continuously. In the context of these recent events and developments the following paper will discuss how the need for storage capacities could be met by the *"power-to-gas"* technology, noting the need for flexibility, close market maturity and economic affordability.

---

[1] http://www.zukunftsenergien.de/hp2/downloads/vortraege/rolle-akz45.pdf S.8
[2] http://www.bundesregierung.de/Content/DE/Artikel/2012/01/2011-01-13-strommarkt.html
[3] http://www.photovoltaik.eu/nachrichten/details/beitrag/erneuerbare-erreichen-erstmals-25-prozent-marke_100008664/
[4] http://www.wind-energie.de/sites/default/files/download/publication/abschaetzung-der-bedeutung-des-einspeisemanagements-nach-ss-11-eeg-und-ss-13-abs2-enwg/20121206_ecofy_studie_einsman_final.pdf

1. Introduction

1.1 Market structure analysis

When looking at the German market structure, like also other countries markets structures, the underlying principle for a society dependent on electricity is the security of supply. The power plants capacity development in the past, and to a big extent also in the future, had to orientate themselves along the annual electricity demand and the maximum demand (peak load). This means that there had to be enough capacities not only to meet the total daily required energy but also the daily time curve of this demand. Therefore, the installed and secured capacity had to be compared to the total demand for electricity and balanced continuously. Especially for the future this will become more difficult, since on one hand the development of the demand and its timing have to be forecasted and on the other hand because renewable energies are classified with a lower secured rate than conventional power plants. In this context, conventional power plants along with deep geothermal and biomass power plants have a secured rate of 85% (losses because of reparation and maintenance) while run-of-the-river hydroelectric, wind and photovoltaic power plants have a secured rate of 50%, 10% and 1% respectively (losses because of lack of water, wind or sun, reparation and maintenance).[5] However, these can still exceed the secured rates by much, creating an excess, as they often do.

The security of supply under the circumstances of an increasing share of the renewable energies will only work when the market design will be changed by the regulator. Otherwise, because of the merit order, the development of the renewable energies will force gas- and coal-fired power plants out of the market way too often for them to be profitable. The introduction of capacity markets, which for example could secure combined cycle power capacities, is therefore very likely. Since these power plants are highly efficient, have a low minimal performance and are highly adjustable,[6] they represent an economical balancing power and hence could compensate for fluctuations in the electricity production of renewables. Thus guaranteeing supply security.

Because of these facts, scenarios for the development of the energy demand, the capacities through renewable energy power plants, the amount of the energy these will supply and the time curve of demand, are essential. According to the already mentioned energy efficiency targets of the Federal Government the total consumption of electricity in 2020 should amount to around 548 TWh (-10% of current consumption of 608.8 TWh (2011)).[7] Following the argumentation of the *"Federal Government"*, according to older plans, the target for the share of renewable energies in the electricity production should exceed 35% by 2020.[8] In the light of recent events, however, the Federal Minister for the Environment, Peter Altmaier, raised this threshold to 40% lately.[9] A called in study from the *"Fraunhofer Institute for Wind Energy and Energy System Technology" (IWES)* predicts that by then 20% would come from wind energy and 10% from solar energy.[10] Recalling the fact that the secured rate of these generation means are very low, the installed quantities have to be very much higher, the excesses have to be stored through storage capacities or the time curve has to alter until 2020.

---

[5] http://www.oeko.de/oekodoc/971/2009-003-de.pdf P. 17f
[6] http://www.siemens.com/press/pool/de/materials/energy/2011-05-irsching4/praesentation-balling-d.pdf
[7] http://www.bmwi.de/BMWi/Redaktion/Binaer/Energiedaten/energietraeger10-stromerzeugungskapazitaeten-bruttostromerzeugung,property=blob,bereich=bmwi,sprache=de,rwb=true.xls
[8] http://www.erneuerbare-energien.de/files/pdfs/allgemein/application/pdf/ee_innovationen_energiezukunft_bf.pdf P. 47
[9] http://www.klima-sucht-schutz.de/mitmachen/beitrag/article/eeg-altmaier-will-grundlegende-reform.html
[10] http://www.erneuerbare-energien.de/files/pdfs/allgemein/application/pdf/ee_innovationen_energiezukunft_bf.pdf P. 47

1.2 Necessity for storage capacities

When prospecting the future, by looking at scenarios about the market design, the load curve and the energy demand and targets about energy efficiency in different sectors, it is inevitable to see that if Germany wants to achieve a complete supply of electricity through renewable energies, storage capacities will be needed. In total, 8% of the yearly electricity consumption could have to be stored at some point[11] which on the basis of the 2011 net consumption of 511.8 TWh[12], would total to a storage capacity of about 41 TWh. With decreasing electricity consumption this figure will decrease as well and find itself somewhere between 20 and 40 TWh.[13] For comparison, the installed capacity of pumped storage hydropower stations amounts to 0.06 TWh.[14] These storage capacities need to be filled-up by the excessive electricity production of renewables which in the future could amount to an overproduction of 80-90 GW.[15]

While the necessity for storage capacities is present, the closure to the market and the economical aspect of possible technologies are still questionable. In this context, it is important to look at benchmark studies to identify which technologies are the most reasonable to use and develop. Mirjam Sucher's seminar paper[16] shows that the power to gas technology could be one of the leading storage systems in the years to come.

1.3 Power to Gas

The power to gas technology, mainly developed by Prof. Dr.-Ing. Michael Sterner (University of Regensburg) and Dr. Jürgen Schmid (Head of the IWES-institute in Kassel), is a method to convert the excessive electricity in the system into the form of electrochemical energy which can then be stored in the gas network, in gasometers or in subterranean caverns. The biggest advantage of this form of storage definitely is the possibility to transport the energy carrier very easily and nearly without restrictions through the whole country. This makes it possible to transform, store and regenerate electricity in completely dispatched locations without making use of the temporarily overloaded electricity grids. Gas from renewable energies will help to decarbonise different sectors, to improve the security of supply in times of electricity grid restrictions and to generate CO_2-neutral fuel. This will decrease dependencies on energy resources and hence could smoothen international relations and conflicts. The process of transformation comprises the step of electrolysis in which hydrogen is being produced and the step of methanation in which the hydrogen reacts with carbon monoxide or dioxide to form methane, essential part of natural gas. These two steps can be looked at independently as both products, hydrogen and methane, can be used and to a certain degree stored in our current energy system. Therefore, the following will analyse them separately as indeed two steps of a process and two differently marketable products.

---

[11] http://www.energieverbraucher.de/de/Umwelt-Politik/Politik/Deutschland/Energiewende__1466/ContentDetail__12975/
[12] http://www.ag-energiebilanzen.de/viewpage.php?idpage=118&preview=true Energieverbrauch in Deutschland 2011 - P. 7.1
[13] http://www.greenpeace-energy.de/fileadmin/docs/sonstiges/Greenpeace_Energy_Gutachten_Windgas_Fraunhofer_Sterner.pdf P. 12
[14] http://dip21.bundestag.de/dip21/btd/17/103/1710314.pdf P. 2
[15] http://www.dw.de/forscher-fordern-schnellen-solarausbau/a-16288089
[16] M. Sucher – Benchmark of storage technologies (2012), P. 13

2. Electrolysis – Step one

2.1 Technological and operational background

Hydrogen is the simplest element in the world and makes up for most of the known matter, however, it is locked up in water, in hydrocarbons (such as methane) or in other organic material. The production of hydrogen (fig. 1) from these components was a big challenge in the past and will be for the future development of more efficient procedures.

Until recently the production of hydrogen was achieved through the direct utilisation of one primary energy source (solar-based, fossil), explaining that hydrogen is an energy carrier and not source. The current developments in the energy market, with fossil fuels predestined to vanish, biomass fully dedicated to other means, and other renewable energies showing big fluctuations in their electricity production, make it necessary to produce hydrogen indirectly through water and a secondary energy source, electricity, and therefore through the process of electrolysis. Additionally, it is essential to make this more efficient for a wide-scale application. In this sense, different procedures are being deployed to split the components of water into hydrogen and oxygen.[17]

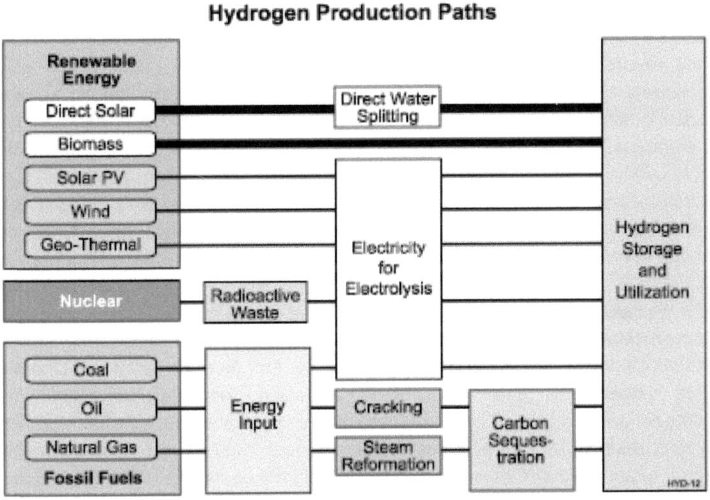

Figure 1

While especially the production through steam reformation (95% of US hydrogen production[18]) or gasification (cracking), fuelled by fossil fuels, is most cost efficient, the costs of electrolysis are twice to three times as high.[19] Moreover, as soon as electricity from renewable resources is being directly employed for the electrolysis process the costs tend to increase as their price is mostly higher. In fact, the costs rise according to the cost developments of renewable energy sources, which show onshore

---

[17] http://www.fsec.ucf.edu/en/consumer/hydrogen/basics/introduction.htm
[18] http://www.afdc.energy.gov/fuels/hydrogen_basics.html
[19] http://www.scribd.com/doc/111500859/Hydrogen-Pathway-Cost-Analysis P. 3

wind power in the lead and solar power considerably more expensive.[20] However, if not analysed in the context of a combined production facility (wind power plant attached to electrolyser) these costs are dependent on the price projection at the energy exchange markets or individual agreements with electricity producers. In this context, and especially through the known price fluctuations at the *"European Energy Exchange" (EEX)* with a generally decreasing trend[21], the costs for the hydrogen generation are volatile.

The history of the electrolytic process goes back to the time around the 18th century when Galvani's and Volta's first experiments were followed by William Nicholson who firstly broke down water into hydrogen and oxygen.[22] "Electrolysis involves passing an electric current through water. The current enters the electrolysis device through the cathode (a negatively charged electrode), passes through the water, and leaves through an anode (a positively charged electrode). Hydrogen is evolved and collected at the cathode and oxygen is generated and collected at the anode."[23] The chemical reaction is defined as **2 H_2O + energy $\leftrightarrow$ 2 H_2 + O_2**. Electrolysers can be hand-sized but research is being committed to increase the scale of electrolytic operational plants which can be integrated at renewable power plants, allowing to shift production to best match resource availability and market factors (demand, price level), or produce big quantities of hydrogen centrally.

There are three electrolytic processes which are being used more frequently, the PEM-Electrolysis, the Alkaline-Electrolysis and the High-Temperature-Electrolysis. "In a polymer electrolyte membrane (PEM) electrolyser, the electrolyte is a solid specialty plastic material. Water reacts at the anode to form oxygen and positively charged hydrogen ions (protons). The electrons flow through an external circuit and the hydrogen ions selectively move across the PEM to the cathode. At the cathode, hydrogen ions combine with electrons from the external circuit to form hydrogen gas." This method is rather new and the best suitable catalysts are still in development as a new catalyst, 97% more cost efficient, shows.[24] "Alkaline electrolysers are similar to PEM electrolysers but use an alkaline solution (of sodium or potassium hydroxide) that acts as the electrolyte. These electrolysers have been commercially available for many years."[25]

Additionally, the process of high-temperature water splitting, which different than a pure electrolytic is a thermochemical process, is a future technology in the early stages of development. "High-temperature heat (500°C–2000°C) drives a series of chemical reactions that produce hydrogen. Chemicals used in the process are reused within each cycle, creating a closed loop that consumes only water and produces hydrogen and oxygen. The high-temperature heat needed can be supplied by next-generation nuclear reactors under development (up to about 1000°C) or by using sunlight with solar concentrators (up to about 2000°C)." However, considering the enormous local and source-specific risk of nuclear energy plants, this option should be discharged by conscious governments. In the last months different alternative procedural ways (more than 200 cycles) have been tested and further research is focused on twelve specific cycles which can improve the efficiency in the future. Further, the

---

[20] COCKROFT, C, & OWEN, A 2007, 'The Economics of Hydrogen Fuel Cell Buses', Economic Record, 83, 263, pp. 359-370, Business Source Premier, EBSCOhost, viewed 28 December 2012. – P. 366
[21] http://www.eex.com/de/Marktdaten/Handelsdaten/Strom/Stundenkontrakte
[22] Iovine, John. "Fuel Cells." Poptronics 2, no. 2 (February 2001): 49. Business Source Premier, EBSCOhost (accessed December 9, 2012).
[23] http://www.fsec.ucf.edu/en/consumer/hydrogen/basics/production-electrolysis.htm
[24] http://www.greenoptimistic.com/2010/05/19/gridshift-electrolysis-catalyst/#.UN4P4ORQWSo
[25] http://www1.eere.energy.gov/hydrogenandfuelcells/production/electro_processes.html

identification of suitable materials for these high temperature operations and the cost reduction of solar concentrators, as well as the development of solar receivers and the applied heat-transfer medium will drive efficiency. High-temperature water splitting is best applicable for large-scale and centralised production of hydrogen and to some extend to semi-central production from solar driven cycles.[26]

When talking about the production cost of hydrogen the main cost driver is the efficiency rate of the utilised electrolytic process. Depending on various parameters, the efficiency rates can vary a lot and are still subject to heavy improvement, as the development of new catalysts, membranes and the research and basic research stadium of the PEM- respectively the High-Temperature-Electrolysis are showing. Different examples of constellations, however, show a range of efficiency between 62-87% for the Alkaline-Electrolyser. While this is the state of art (2011) the mid-term (5-10 years) and long-term (10-20 years) potential is given as ranges from 67-82% respectively from 67-87%.[27] The PEM-electrolyser is showing a current efficiency range of 67-82% and future ranges from 74-87% and 82-93% respectively.[28] It has to be considered that the higher range values are absolute maximums which are not reached consistently and very often. High-Temperature-Electrolysers are still in development and their efficiency will be predominantly defined by the current density and the cell voltage.[29] The variable costs for an electrolyser, therefore, will be calculated by the attitudinal efficiency multiplied by the electricity price which can be secured on the volatile market. Taking into consideration this year Christmas Day's hourly spot market prices for the German market, one kWh at the EEX cost about 3 Eurocents and at night purchasers even received money for taking off load.

An exemplary calculation with an efficiency rate of 75%[30] would result that the equivalent of 1 kWh of hydrogen would require approximately 1.33 kWh of electricity and hence approximately 4 Eurocents. While this figure looks lower than the price for natural gas, other costs have to be integrated before finding the real end cost. These comprise the amortisation cost of building the plant, operational and insurance cost, maintenance and wages, moreover the transportation costs have to be evaluated. In a study[31] from 2008 these costs are being identified to amount approximately 1.31-1.75 Eurocent/kWh depending on which technology is being used. Finally, a profit margin has to be added (20%) to the costs, resulting in a reasonable end price of 6.5-8 Eurocent/kWh.

Concluding, it has to be said that electrolysers are sensitive to load fluctuations. With increasing efficiency the possible partial load decreases and shows values from 40-10% for the alkaline process and 10-0% for the PEM-process.[32] A steady and secure supply, even if at higher prices, has to be ascertained and accounted for (battery, flywheel).

---

[26]

[27] http://www.now-gmbh.de/fileadmin/user_upload/RE-Mediathek/RE_Publikationen_NOW/NOW-Studie-Wasserelektrolyse-2011.pdf P. 30

[28] http://www.now-gmbh.de/fileadmin/user_upload/RE-Mediathek/RE_Publikationen_NOW/NOW-Studie-Wasserelektrolyse-2011.pdf P. 31

[29] http://www.now-gmbh.de/fileadmin/user_upload/RE-Mediathek/RE_Publikationen_NOW/NOW-Studie-Wasserelektrolyse-2011.pdf P. 16

[30] http://www.zeit.de/angebote/zukunftswerkstatt/erdgas-2/enormes-potenzial/index

[31] http://www.ika.rwth-aachen.de/r2h/index.php?title=Hydrogen_Pathway:_Cost_Analysis&oldid=5029

[32] See sources 25 and 26

2.2 Origin of electricity

As mentioned in the introduction the development of the renewable energies will continue to boost in the next years and reach a considerable share of over 40% by 2020. Recalling the secured capacity of the renewable electricity sources there will still be a lot of parallel production to secure the electricity supply in Germany. This, on one side, is essential for our society and industry, on the other side, however, implies giant surpluses in the production. Additionally, the hampering development of the electricity grid and the inability up to know to store and transport electricity will cause a lot of capacities to be forced to shut down. Indeed, last year 407 GWh of wind power got lost due to these restrictions. The scope of the electrolysis and the further steps of the power-to-gas technology is to store exactly these otherwise lost units of electricity to enable them to be consumed in another period. This would increase the secured amount of renewable energy production and could hence increase their share in the total energy mix. While 407 GWh of input is relatively few for the profitable storage via the power-to-gas technology, the doubling of the wind power capacities and the quadrupling of the photovoltaic capacities[33] will increase this figure considerably. Additionally, the inflexibility of our current energy system will cause enormous surpluses through the merit order effect. This means that in times where the favoured renewables are strong in production and hence consumption other generation means will be pushed out of the market. In such a case, the operators of these plants need to export their electricity or downgrade their operation. Since the load gradient and the minimum load of coal and nuclear power stations do not permit this quick response electricity net surpluses of a total of nearly 145 TWh of electricity (surpluses of 190 TWh – deficits of 45 TWh) per year are created (residual load).[34] These surpluses can be transformed and re-electrified when deficits make it necessary (fig. 4).

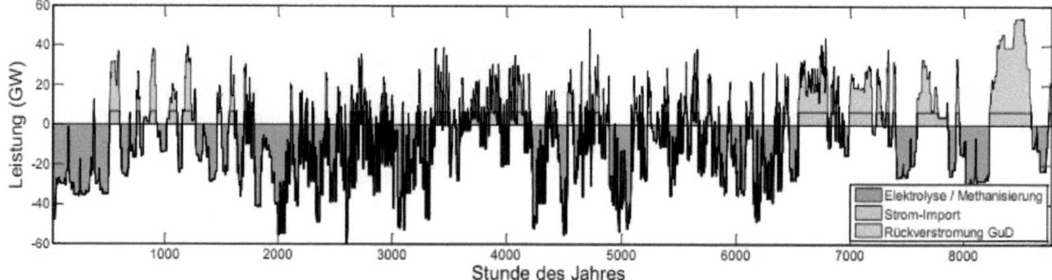

Figure 2

In such a scenario it has to be considered that up to now the limited capacities (shut off because of feeding-in management) are getting a compensation (at least the renewable sources) according to the *"Renewable Energies Act"* (§ 12/13) which amounts to 90% of capacity utilisation. In future, these compensations would need to be shifted to the successful conversion of the surplus energy into stored energy. A possibility could be a splitting of this compensation (paid over the EEG reallocation charge by all consumers) between the producer and the conversion company. This would decrease the profitability of renewable energy plants but improve it on the other hand for power-to-gas operating companies, which again could motivate operators of for example wind power plants to engage also in the business of conversion and hence profit from synergies.

---

[33] http://www.zsw-bw.de/fileadmin/ZSW_files/Infoportal/Presseinformationen/docs/Vortrag_IWES_Schmid_Einweihung_Methanisierung.pdf P. 2
[34] See source 33, P. 4

2.3 Potential and utilisation of hydrogen

Since hydrogen has a minimal environmental impact if it is being used, and it can replace fossil fuels, it is an interesting resource for the future. While it is already being used in space for power and oxygen supply and to store drinkable water for the astronauts[35], a less futuristic scenario could be the adoption in the automobile sector. While the development of fuel cells and their utilisation in cars is actually already technologically implemented as can be seen by the releases of BMW with the 1 series-fuel cell hybrid (2010), Mercedes-Benz with the F800 (2010) and the B-Class-F-Cell (2009), Fiat with the Panda-HyTRAN (2009), Volkswagen with the Caddy-Maxi HyMotion (2009) or Audi with the Q5-FCEV (2009) but also various busses like the Mercedes-Benz fuel cell bus, issues with the on-board storage and the network of power stations, as well as the fuel cost are preventing the breakthrough. In fact, even when compressed at nearly 700 bar (gasometers usually have a pressure of 5-20 bar) the size of a tank permitting a range of 500km is three times as big as a gasoline tank and even liquefied hydrogen would need a tank twice as big.[36] However, research continuous and the programme: *"Targets for Onboard Hydrogen Storage Systems for Light-Duty Vehicles"* from the *"US Department of Energy Office Energy Efficiency and Renewable* Energy" *(DOE)* is targeting to equal their size by 2015.

Additionally, next to the size issues, there are many other issues which have to be solved in order to use fuel cells and hydrogen on a wide scale. These include "operating pressure and temperature, the life span of the storage material, the requirements for hydrogen purity imposed by the fuel cell, the reversibility of hydrogen uptake and release, the refuelling conditions of rate and time, the hydrogen delivery pressure,"[37] toxicity and overall safety imposed by high pressure and low temperatures (liquefaction). Many of these attributes seem not to be achievable via gaseous hydrogen but rather through its absorption by other chemical materials. The programme is trying to tackle these last hurdles on the way to commercialisation of hydrogen fuelled vehicles.[38] Moreover, the U.S. government has planned several budgets to develop the linked technologies to reach commercially viable hydrogen-powered fuel cells by 2020.[39] And also in Europe the need for intensive research in the form of public private partnerships and with the application of governmental incentives is being recognised, even though only investing about half than the US.[40] Examples are the HyFLEET:CUTE- or the ITHER-project.[41] At the current state, however, the cost, volume and weight of the three different adopted and researched modalities (gaseous, liquid, solid) is shown in figure 2.

---

[35] http://www.nasa.gov/topics/technology/hydrogen/hydrogen_2009.html
[36] http://www.afdc.energy.gov/fuels/hydrogen_basics.html
[37] http://www.fsec.ucf.edu/en/consumer/hydrogen/basics/storage.htm
[38] http://www1.eere.energy.gov/hydrogenandfuelcells/storage/pdfs/targets_onboard_hydro_storage_explanation.pdf
[39] http://www.hydrogen.energy.gov/pdfs/hydrogen_posture_plan_dec06.pdf P. ii
[40] http://www.gppq.mctes.pt/_7pq/_docs/brochuras/online/111026_20FCH_20technologies_20in_20Europe_20-_20Financial_20and_20technology_20outlook_202014_20-_202020.pdf P. 8,9,19
[41] http://www.now-gmbh.de/fileadmin/user_upload/RE-Mediathek/RE_Publikationen_NOW/NOW-Studie-Wasserelektrolyse-2011.pdf P. 26

Storage Technologies	Weight (kwh/kg)	Volume (kwh/L)	Cost ($/kwh)
Chemical Hydrides	1.6	1.4	$8
Complex Metal Hydrides	0.8	0.6	$16
Liquid Hydrogen	2.0	1.6	$6
10,000-psi Gas	1.9	1.3	$16
DOE Goals (2015)	3.0	2.7	$2

Figure 3

The storage and distribution of bigger quantities of hydrogen, however, is not technologically constrained anymore. Hydrogen can be compressed to approximately 685,000 litres per square metre[42] and transported by pipelines (specific and combined with natural gas), high-pressure tube trailers (low distances), in liquefied hydrogen tankers (long distances) or in solid state either by absorbing or reacting with metals or chemical compounds.[43] However, the investments into infrastructure, especially for an extensive distribution network, are high, since own pipelines would have to be built throughout the country or tankers (rail and roadway) would need to constantly move. Even though hydrogen can be produced in a decentralised way and hence reduce the absolute need for transportation, there are efficiency trade-offs which need to be weighed against distribution costs.

2.4 Greenpeace Energy – proWindgas

Greenpeace Energy is a German utility cooperative founded in 1999 to provide electricity free from coal, natural gas and nuclear power. By today it has over 110.000 customers and 22.000 comrades. Since 2011, a gas tariff is being offered which in future will supply customers with hydrogen produced by excessive load of the own eight wind parks and three photovoltaic plants (further to come) with a total capacity of 54 MW.[44] The usage of these surpluses in production through the generation of hydrogen stabilises the grid while the gas anon can be easily stored in the natural gas grid covering the whole country. In fact, the blending of hydrogen in methane is not a problem for up to 5% in a technological and regulatory sense[45] and could even be extended to 10%[46] or more (however, decreasing the energy value). In this way, the enrichment of the methane resources in the natural gas grid reaches consumers who can use the gas for cooking, charging their automobile or to generate heat and electricity in a cogeneration unit. By the end of 2013 the electrolyser will take up its operation and will mainly be financed by the comrades and the amount of funding paid by the customers up to date. Indeed the charged price of 7.51 Eurocent/kWh (similar to the one calculated at the end of section 2.1) includes 0.4 Eurocent which are dedicated to the development and construction of the electrolyser and the total system technology.[47] A generic pioneer in a future technology which is arousing interest also in government circles.[48]

---

[42] http://www.afdc.energy.gov/fuels/hydrogen_basics.html
[43] http://www.afdc.energy.gov/fuels/hydrogen_production.html
[44] http://www.presseportal.de/pm/16698/2328391/neuer-schwung-fuer-die-energiewende-windgas-made-in-suderburg-greenpeace-energy-will-in
[45] http://www.greenpeace-energy.de/engagement/unsere-gasqualitaet/windgas-als-speicher.html
[46] http://www.zeit.de/angebote/zukunftswerkstatt/erdgas-2/enormes-potenzial/index
[47] http://www.greenpeace-energy.de/windgas.html
[48] http://www.presseportal.de/pm/16698/2320435/bundesregierung-haelt-windgas-fuer-besonders-interessant-windgas-steht-im-10-punkte-programm-des

3. Methanation – Step two

3.1 Technological and operational background

To avoid the need for an exclusive storage and distribution system and to reach a (three times) higher energy density in gaseous state (kWh/Nm^3)[49] the process can go one step further and create synthetic natural gas by making hydrogen react with carbon dioxide (but also monoxide). This process is called a Sabatier reaction after the French chemist Paul Sabatier who published his work about hydrogenation and catalysts in 1913. The reaction defines as **CO_2 + 4H_2 → CH_4 + 2H_2O** and ensues exothermically generating considerable amounts of heat. Different than the predominantly electric process in the electrolytic process the methanation process requires a high temperature and pressure (activation energy) as well as a high CO2 concentration and purity. Key to the efficiency of this procedure is the used catalysts which enhance the speed and required energy of the reaction and the constant and consistent heat dissipation.[50] In future the conception of the reactors and the reaction operational procedure need to be improved, to accomplish these premises.[51]

The known operational plants are classified as two-phasic (fixed-bed, fluidised-bed) and three-phasic (bubble column) reactors (fig. 4), both operate with gaseous reagents and solid catalyst and for the three-phasic reactor a heat transfer medium is being applied additionally. However, while latter are still in the stage of development and are promising to be less sensitive to load fluctuations, currently especially fixed-bed reactors are being used.[52] In a fixed-bed reactor the catalyst is present in the form of pellets which are held in place and are only able to move to a certain degree. Since the input gases have to pass on the catalysts to react, the capacity is limited to the materials surface. As well, the reactor need to be cooled to prevent the catalysts to become obsolete.[53] Catalytic fixed-bed reactors are the most used when it comes to large-scale blending of chemicals.[54] Opposed to them, fluidised-bed reactors still contain solid catalyst which, however, move freely in the reactor. By this, they allow the gas to access the catalyst better allowing them to flow over a greater share of the surface. Additionally, heat issues are amended by the circulation of the catalyst material. Finally, catalysts can be removed from the reactor for regeneration or replacement without having to shut it down.[55]

A bubble column reactor consists of vertically arranged cylindrical columns including the catalyst material. The introduction of gas at the bottom into a liquid-solid suspension causes turbulent streams and bubbles enhancing an optimum gas exchange and at the same time saving energy for the mixing.[56] Additional benefits are the excellent heat and mass transfer characteristics as well as little maintenance and low operating costs. Current research is being performed on issues as bubble characteristics, flow dynamic, heat transfer mediums and the overall conception, temperature and pressure. Even if a lot of research is being conducted, bubble column reactors are still in development since these studies manly have to constrain to focus only on one phase (liquid or gaseous) and not their inevitable linkage.[57]

---

[49] http://www.h2data.de/
[50] http://www.powertogas.info/power-to-gas/strom-in-gas-umwandeln.html#c170
[51] http://www.powertogas.info/power-to-gas/strom-in-gas-umwandeln.html#c170
[52] http://www.powertogas.info/power-to-gas/strom-in-gas-umwandeln.html#c170
[53] http://www.netl.doe.gov/technologies/coalpower/gasification/gasifipedia/5-support/5-14_gas-catalysts.html
[54] http://elib.uni-stuttgart.de/opus/volltexte/2009/4798/pdf/eig16.pdf P. 200
[55] http://www.netl.doe.gov/technologies/coalpower/gasification/gasifipedia/5-support/5-14_gas-catalysts.html
[56] Bubble Column Reactions, by Wolf-Dieter Deckwer, John Wiley & Sons, ISBN 978-0-471-91811-0
[57] http://web.ist.utl.pt/ist11061/de/Equipamento/BubbleColumnReactors(review).pdf P. 2264

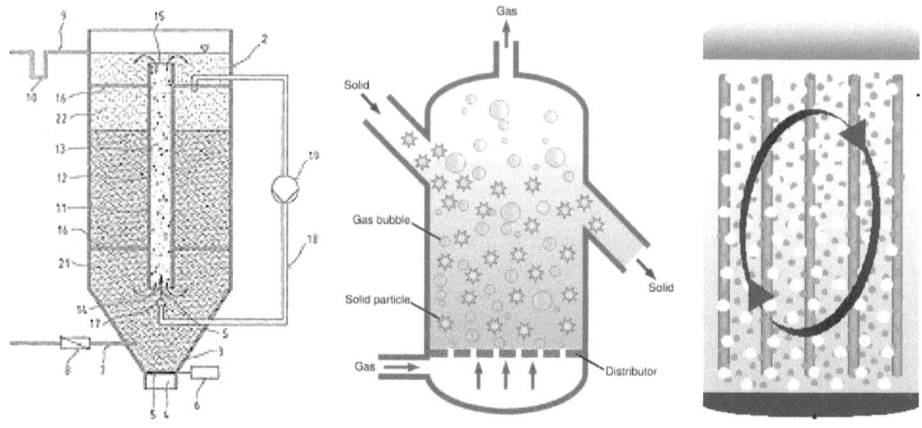

Figure 4

The methanation procedure decreases the efficiency of the hydrogen production effectiveness by another 12-15%. [58] Depending on the different applied technologies and framework conditions (pressure, temperature) the system efficiency of the power-to-gas technology is therefore in the range of 49-65%.[59] In the least efficient way, the production of one unit of synthetic methane hence requires two units of secondary energy (electricity) and by considering the efficiency of wind power plants (approx. 50%) or photovoltaic modules (15%), four respectively over 13 units of primary energy. Taking into consideration an average price of 7.5 Eurocent/kWh of hydrogen and putting it into relation with average system efficiency rate of 55% (resulting through a methanation efficiency of approximately 75%), the resource input cost for the generation of synthetic methane amounts to approximately 10 Eurocent/kWh excluding the additional costs for land, labour and capital. However, the initial investment is estimated at €1bn per GW of capacity[60]. The low efficiency rates, though, have to be put into comparison to procedures which would nonetheless occur, and hence the complete loss of the potential energy capturing. The sun is constantly sending energy to the earth and nature is only saving less than 1% of it in chemical energy carriers (biomass, natural gas, oil). The capturing of the sun energy through photovoltaic panels and its storage through the power-to-gas technology would hence increase the utilisation rate twentyfold and more compared to biogas or -mass.[61] This shows that even though the process of methanation increases the conversional work even more and total efficiencies are comparably low, this technology is an important stepping stone on the way to a more renewable energy system permitting a higher yield.

When electrifying the produced synthetic methane the total system efficiency amounts to 30-38% and when using combined heat and power processes to 43-54% [62] (process efficiency of 58-61%

---

[58] http://www.zeit.de/angebote/zukunftswerkstatt/erdgas-2/enormes-potenzial/index
[59] http://www.greenpeace-energy.de/fileadmin/docs/sonstiges/Greenpeace_Energy_Gutachten_Windgas_Fraunhofer_Sterner.pdf P. 18
[60] http://www.energieverbraucher.de/de/Umwelt-Politik/Politik/Deutschland/Energiewende_1466/ContentDetail_12975/
[61] See source 55, P. 18
[62] See source 55, P. 18

respectively 83-88%). This means that three kWh of electricity need to traverse the process to make it possible that one kWh can be supplied in a chronologically offset period. Setting this into relation with the calculated cost of synthetic methane (10 Eurocent/kWh) and efficiency rates of the previous steps, it would result in a cost of approximately 16.5 Eurocent/kWh$_{el}$ and approximately 12 Eurocent/kWh$_{th/el}$. Since the generated heat is a valuable commodity it should be used and therefore nearby purchasers (companies, district heating grid) These costs are again not including further costs but only represent an estimation of the costs. Especially the profit margin applied in the calculation for the hydrogen production would not be applied on every step but only on the last (if same company).

3.2 Storage and transportation

The biggest advantage of the methanation process is the fact that the generation of synthetic methane exploits the whole potential of the natural gas grid for storage and transportation. Even if the German electricity grid would be perfectly enlarged and combined and integrated into the European transmission grid, surpluses would amount up to 170 TWh$_{el}$ which cannot be dealt with in sync.[63] Indeed, the input power and the pipeline/grid capacity differ considerably for electricity and natural gas. While the electric transmission grid permits only one-digit GW amounts (around 3 GW) the natural gas transmission pipelines permit up to 70 GW.[64] This notable difference reflects the advantage of the power-to-gas technology which allows to store the excessive energy but especially to transport the energy in a much wider scale according to the geographically circumstances affecting the electric yield of renewable power plants and the required load affecting local generation of electricity. In fact, Germany disposes for nearly 450 000km of transmission and distribution grids[65] enabling the generated synthetic methane to reach every part of the country.

The natural gas grid in Germany accounts for a storage capacity of 220 TWh$_{th}$ which at efficiency rates of about 55% (combined cycle power plants) achieves 120 TWh$_{el}$ when electrified again. These represent 2 to 3 months of Germany's electricity consumption which in gas terms totals at 900 TWh$_{th}$.[66] Moreover, the German gas grid alone represents for a total of 23.5 billion m^3 of aboveground and underground gas reservoirs, the biggest in Europe.[67] Additionally, the expansion of the gas grid, caverns and pore storage will double in the decades to come and the sustainable maximum amounts to 514 TWh$_{th}$ (280 TWh$_{el}$)[68] This shows how incredible flexible and extensive an already available energy grid is and that the *"Electricity Grid Development Plan"* of the *"Federal Network Agency"* and the transmission grid operators is completely unnecessary in the planned scale. The development and usage of the power to gas technology would be a better investment for a reasonable share of the budgeted €20bn.[69] This is also more sensible in the long term when considering the losses the different grids are experiencing. An AC-transmission grid shows a loss of up to 10% per 1000 km while natural gas transmission grids only lose 0,5% per 1000 km (in an ecological perspective the leakage of methane is however more harmful than the emission of CO2). Moreover, electricity grids are dimensioned for transporting the peak energy amount. In non-peak times the capacities of the grid are

---

[63] See source 55, P. 8
[64] See source 55, P. 10
[65] http://www.zeit.de/wirtschaft/2011-09/power-to-gas/komplettansicht
[66] See source 55, P. 16
[67] http://www.dena.de/fileadmin/user_upload/Projekte/Erneuerbare/Dokumente/PowertoGas_Thesenpapier_Technik.pdf P. 2
[68] See source 55, P. 17
[69] http://www.wiwo.de/politik/deutschland/energiewende-netzausbau-kostet-20-milliarden/6690132.html

underutilised at about 40-60%.[70] The power-to-gas would make it possible to reduce the amount of grid capacity by substituting the transmission network by the gas network. Through this, the placement of stored energy and small generation units next to varying demand (industrial zones) can cancel out peak situations and allow a smoother and more equal supply fuelled by renewables.

3.3 Potential and utilisation of synthetic methane

For our current society and following generations carbon dioxide is a troublesome gas which however, originating from coal, is part of the most life-essential material. The trade-off between our consumption-based society and the pollution of our environment has to be managed and the effects decreased to maintain our basis of existence. Through the power-to-gas technology the current most pressing issue and problem of a simultaneous generation and consumption of electricity would be solved. This can be said as such by the fact that the generation does not have to be planned and coordinate upon secured capacities anymore but can be managed in a more flexible way by providing a base load and filling up fluctuations of renewable energy sources nearly concurrent by the utilisation of the synthetic gas in gas and combined cycle power plants and cogeneration units. Moreover, the interchangeability of the synthetic methane for different markets, makes this energy carrier a perfect element for the general energy turnover. Synthetic methane can be used for electricity generation as well as for heating, cooking and for mobility.[71] Additionally, the up-scaling of the usage of synthetic methane in the different markets is not showing any difficulties since the technology and mostly also the infrastructure is available in all of these and can be gradually expanded (combined cycle power plants, cogeneration units, gas-powered vehicles)[72]

Especially for the mobility sector, the most polluting after the power generation, synthetic gas could be the easiest option of those being examined today. The technology is pretty developed, the tanks large enough in size to permit a considerable range and the fuel distribution system is already present with many fuel stations giving the possibility to fill the tank quickly and new ones easy to install. The efficiency to generate synthetic methane is promising and combined with the yield of biogas plants and methane future production possibilities (algae) the technology could reduce the carbon dioxide emissions, the personal costs for mobility and the dependency from oil-exporting countries.

In future, the main trouble in an energy system with a high stake of fluctuating renewable energies will not only be the restrictions of the electricity grid but more the alternation of long phases of high overproduction (abundant wind power in autumn or photovoltaic in spring) and phases of lack of production (winter).[73] The only known long-term storage, next to the restricted capacities of pumped storage hydro power stations, can only exist in the form of chemical energy and hence for example synthetic gas. Above all, the energy turnover in Germany has to meet high supply security standards and the ability to react very quickly to load shifts and external factors such as the location of the sun and local spontaneous weather changes. The utilisation of gas fired power plants is ideal for such a system since the starting timeframe for these is low (15 min.) and the load gradients relatively high.[74] A

[70] http://www.itas.fzk.de/deu/lit/epp/2011/pasc11-pre01.pdf P.12
[71] http://www.zeit.de/2011/37/Speicherung-Oekostrom
[72] http://www.zdf.de/ZDFmediathek/beitrag/video/1658560/Projekt-Windgas#/beitrag/video/1658560/Projekt-Windgas
[73] http://www.zeit.de/2011/37/Speicherung-Oekostrom
[74] http://www.alt.fh-aachen.de/downloads/Vorlesung%20EV/Hilfsb%2060%20Regelleistungsbereiche%20Lastgradienten%20Kraftwerke.pdf P. 2

market design of capacity markets for gas and combined cycle power fuelled by synthetic methane could help these type of power plants to reach profitability again. In the past, the news about the shutting down of these were quite recent[75] and hence such a policy could kill to bird with one stone. Additionally, the low scale application of the gas technology (e.g. cogeneration units) makes a widespread and hence decentralised generation possible.

3.4 Origin of CO_2

The methanisation can take place with the reactant carbon dioxide originating from fossil (coal power plants, calk or cement production) or renewable sources (biogas and sewage plants). Additionally, it can also be absorbed from the air. Considering the decarbonisation strategy of the *"Federal Government"*[76] renewable carbon dioxide sources should be favoured and synergies should be stipulated. Nevertheless, as soon as the technology is being implemented in a large scale, the amount of carbon dioxide from regenerative sources will become rare. 20 to 30 million tons of carbon dioxide alone are needed to realise a supply secured provision of storage systems in Germany.[77] Additional tons will be needed if the heating and mobility market will require higher shares of synthetic methane.

Possible synergies could be achieved by connecting methanation reactors to sewage or to biogas plants. Former constellation is being researched by the university of Emden/Leer which found out that the usage of the incurred digester gas, including carbon dioxide and methane can dispose the gas from the operational plants and increase the efficiency and yield of methanation reactors. A second interesting step is the investigation if demand response is possible for sewage plants.[78] Similar to this, the integration of methanation reactors and biogas plants is favourable in two contexts. First, the generation of biogas produces carbon dioxide as a by-product which has to be extracted before feeding in the methane. This process can be bypassed by blending the crude biogas (or only the carbon dioxide) with hydrogen and hence increasing the efficiency of the biogas plant. When combined with a generator a closed system can be established since the water occurring in the combustion can be reused by an electrolyser to start the process all over again. Second, the joined feeding-in of the methane prevents higher leakages.[79]

The carbon dioxide absorption from the air is technically possible, however, as long as sufficient biogenic carbon dioxide is present this option remains too expensive and evitable. In a scenario of missing biogenic carbon dioxide (remote development of wind or solar power) or to decrease the carbon dioxide concentration in the air, and hence to contain emissions of mobile polluters (cars, planes), the technology could be applied and developed.[80] An inexpensive option to recover carbon dioxide is the capturing of it in industrial processes. Similar, the carbon dioxide can also be obtained by using that of carbon capture and storage (CCS) systems applied in the generation of electricity through fossil power plants. However, it has to be considered that the carbon dioxide emissions will not be diminished but only deferred. Latter would also happen to the progress of the energy turnover as the

---

[75] http://www.iwr.de/news.php?id=21200
[76] http://www.bundesregierung.de/Content/DE/_Anlagen/2012/02/energiekonzept-final.pdf?__blob=publicationFile P. 31
[77] http://www.zeit.de/wirtschaft/2011-09/power-to-gas/komplettansicht
[78] http://www.hs-emden-leer.de/aktuelles-termine/news/article/klaeranlagen-als-energiespeicher.html
[79] http://www.greenpeace-energy.de/fileadmin/docs/sonstiges/Greenpeace_Energy_Gutachten_Windgas_Fraunhofer_Sterner.pdf P. 24
[80] See source 72, P. 25

former waste for which companies have to pay (EU ETS 2013-2020) becomes a resource.[81] The best and most sustainable option would be the creation of a recycling system between the synthetic methane generation and the recovery of the carbon monoxide emitted through the combustion of it.

3.6 Solar Fuel Technology

Based on the experiments of the first and second pilot plants (25 and 250 kW) in Stuttgart, the *"Solar Fuel GmbH"* together with its research partners *"Centre for Solar Energy and Hydrogen Research Baden-Württemberg" (ZWS)* and the *"Fraunhofer Institute for Wind Energy and Energy System Technology" (IWES)* will take into operation the first large scale electrolysis and methanation plant (6 MW) in the beginning of 2013. The principal, *"Audi"*, is planning to produce 3,900 m^3 of synthetic gas per day which will be feed into the grid and hence balance the fuel demand of its customers using the gas fuelled Audi A3 TCNG. The conversion plant will have a total system efficiency of 54% (using also the heat discharge) and will receive the carbon dioxide from a biogas plant.[82] These three pilot plants are showing *"Solar Fuel"* that the application of the technology is working and that especially the regulation electronics is already ready for larger plants. The new data from the 6 MW plant will show in which form and under which conditions this up-scaling can take place in the most efficient way.[83]

---

[81] See source 72, P. 26
[82] http://www.powertogas.info/fileadmin/user_upload/downloads/Broschuere/Fachbroschuere_Power_to_Gas.pdf P. 15
[83] http://www.maschinenmarkt.vogel.de/themenkanaele/erneuerbareenergien/articles/383941/

4. Conclusion

It has to be said that this seminar paper should be understood as a general overview over the technology, its environment and the current and future developments of the energy system in which this would be embedded. The clustering of information is a good technique to discover new cognitive paths and to build synergies, however, it should not be forgotten that this also implies inaccuracy and inconsistency. Empirical data, technical knowledge and a more detailed analyse of the contributing singular factors would be necessary to create a more accurate paper.

4.1 Flexibility

As mentioned several times, the power-to-gas technology would permit the energy system to become more flexible. The storage of electrical capacities in the enormous natural gas grid and its extensive network development, will make it possible to react to differences in the current electricity generation and consumption locally and specifically. This will make it possible to gradually downgrade big power plants and replace these by renewable power plants whose production will be stabilised by the power-to-gas and the most efficient gas combustion technologies. Moreover, the ability to transfer the synthetic methane also over the national borders very easily will make it possible to secure the supply and the stabilisation of electricity generation also in nearby countries in a more easy manner than before. Joined projects in certain European regions do not only generate return for the non-local companies but also a product which can be offered in the local market.

4.2 Market maturity

As so often the market maturity will be mainly influenced by political factors and decisions. The power-to-gas technology is readily available as considering it theory and practical application. The up-scaling of the operational plants and further research to optimise efficiency rates only lies in the decision if the technology will be subsidised in terms of research budgets and if the energy system will make use of it. While former is occurring through a budget of €200m for all storage technologies[84], latter, will depend on the future market design since the power-to-gas technology provides a synthetic gas which in a market coordinated according the merit order principle could not compete at the moment, at least from a market-efficiency perspective. A perspective looking at the economy as a whole, however, would come to the conclusion that the advantages of flexibility and mostly renewable origin combined with the essentiality to transform the energy system outweigh the fact of market efficiency. Therefore, the scenario of capacity markets for stored electricity as well as for gas and combined cycle power plants or a feed-in compensation could boost the technology into market maturity in only few years. Since the technology will reach market efficiency quite quickly, the subsidies from the state (or the consumers through a reallocation charge) should be digressive to prevent over-subsidisation as happened in the solar energy market. And indeed, according to the half-governmental German Energy Agency (Dena), this market maturity could already be reached by 2020.[85] According to the previously mentioned Solar Fuel GmbH this point could be meet even before, in 2015. In one way or the other, Germany will

---

[84] http://www.bmwi.de/DE/Presse/pressemitteilungen,did=390582.html
[85] http://www.dena.de/fileadmin/user_upload/Projekte/Erneuerbare/Dokumente/Eckpunkte_Roadmap_Power_to_Gas.pdf P. 2

become leader in the management of the whole systems technology considering its widespread engagement up to now.[86]

4.3 Affordability

In this aspect, the main focus of research will be put on the catalyst to boost efficiency. The developments of cheaper and more efficient catalysts will decrease the cost in sense of refurbishment and by decreasing the time the input materials spend in the electrolyser/reactor as well as increasing overall efficiency.[87] Additionally, as said in section 4.2 the market design is crucial for the profitability of the technology. While in the past peak load prices were high, the current high share of renewables is smoothening the prices, eliminating the economic basis for storage options. Since investments into the electrolysers and methanation plants are high the profitability has to be meet by having enough operating hours. The electricity generation, however, will depend on fluctuations and hence not be source of enough revenue. An operator, therefore, has to prospect different markets for his product of synthetic methane, like the gas, heating and mobility market, and in future maybe even the control energy market.

However, through the revision of the *"Renewable Energies Act"* in 2012 and the *"Energy Economy Law"* in 2011, the power-to-gas operation has become some kind of framework policy which guarantees some financial security. Indeed, since August 2011 all the power-to-gas plants going into operation in the following 15 years will not have to pay grid fees for 20 years. Moreover, if the input electricity derives from renewable sources the output electricity will get a feed-in compensation according based on the initial energy source.[88]

4.4 Outlook

To achieve a complete regenerative provision of electricity, long-term storage possibilities are inevitable. The process of transformation of electricity into synthetic methane is the solution to the problems of electricity supply providing all the gas grid as storage capacity, but also to decrease dependencies on energy imports especially by finding new possibilities to fuel mobility. In the next five years, research will increase efficiencies and plant life-expectancies, will modulise the conversion plants and standardise the operations. And combined with a new market design this storage possibility and affordable module technology will boost the further energy turnover. Additionally, as with every technology finding greater application, new technology branches and interconnections will be developed. One of these could be a technology directly generating hydrogen within a photovoltaic cell using rust instead of expensive semiconductor material. Currently, the efficiencies are low, but in future this technology could help to generate hydrogen nearly everywhere.[89]

The power-to-gas technology will definitely play a major role in the energy systems of the future and hence is worth to be considered and analysed more in detail.

---

[86] http://www.zeit.de/angebote/zukunftswerkstatt/erdgas-2/enormes-potenzial/index
[87] http://www.mpg.de/5597354/kohlendioxid_methanolsynthese_energiespeicher?print=yes
[88] http://www.dena.de/fileadmin/user_upload/Projekte/Erneuerbare/Dokumente/120607_Thesenpapier_Wirtschaftlichkeit.pdf P. 3
[89] http://www.sonnenseite.com/index.php?pageID=6&article:oid=a23828

YOUR KNOWLEDGE HAS VALUE

- We will publish your bachelor's and
  master's thesis, essays and papers

- Your own eBook and book -
  sold worldwide in all relevant shops

- Earn money with each sale

Upload your text at www.GRIN.com
and publish for free